Land and Estate Management

FORESTRY AND WOODLAND MANAGEMENT ON FARMS AND RURAL ESTATES

David Lewis
Royal Agricultural University
Cirencester

LIVERPOOL UNIVERSITY PRESS

FORESTRY AND WOODLAND MANAGEMENT

First published in the United Kingdom, 2021, by Packard Publishing Ltd, 14 Guilden Road, Chichester, West Sussex, PO19 7LA, UK.

This short book corresponds to Chapter 3 of the fifth edition of *Land and Estate Management,* 2021.

ISBN 978 1 85341 167 0 (paperback)

A CIP catalogue record of this book is available from the British Library.

Prepared for press by Michael Packard.

Cover and title-page photo courtesy of Carter Jonas.

Design and Layout by Hilite Design, Marchwood, Southampton, Hampshire.

Printed and bound by CPI Group (UK) Ltd, Croydon CR0 4YY

Contents

Acknowledgements

I am grateful to those people who kindly let me use their photographs in this book: John Lockhart (Figure 3), Lockhart Garratt (Figures 1, 2, 5), and Scottish Woodlands (Figure 7). The other images are my own.

I would also like to thank Richard Scholfield of woodlands.co.uk for his comments on an early draft of the text, and the students at Harper Adams University and the Royal Agricultural University who attended my various lectures and woodland site visits over the last 25 years. It was their interest and perceptions of trees and woodlands, and their observations and questions, which helped shape what to include and what to leave out in the narrative of this introductory book.

Last, I must thank my mother for igniting my interest in plants and trees from an early age, and for supporting me in my choice of career.

D.B.L

INTRODUCTION

The United Kingdom's 3.2 million hectares of woods and forests are made up of a wide range of woodland types, ages and sizes. In turn there are many woodland owners who have different managerial aims that extend from optimizing timber production and financial returns to the management for non-timber benefits, such as wildlife, landscape and recreational enhancement. Furthermore, the growing of timber and the management of woodland is a long-term venture. The intervals between establishing trees, managing and felling them are long. The cycle of production or rotation will frequently take over 50 years for a conifer plantation, and can exceed 100 years for deciduous broadleaves. In addition, trees can survive many hundreds of years. Thus it is important to have a clear, long-term vision for any woodland and to determine the management objectives and strategy for its existence.

FORESTRY AS A LAND USE

Woodland occupies more than 30 per cent of the terrestrial surfaces of both the world and the European countries. In comparison, the UK is relatively sparsely wooded, with only 13 per cent of its land area made up of woodland and forests. At one time much of the land was wooded: the result of the retreat of the ice sheet after the last ice age when the climate became warmer and plants gradually returned, culminating in woodland. Over the years much of that woodland has been felled, though the timing and pattern of the clearance varied considerably in different parts and regions of the UK. For example, the South Downs in Sussex were initially cleared of large swathes of woodland in Neolithic times, whereas the Highland area of Scotland retained most of its woodland well into the eighteenth century. Agriculture was the most common cause for the clearance of woodland in response to the development of settled communities and an increasing population. The historical low point for woodland cover was at the start of the twentieth century, which led to the formation of the Forestry Commission in 1919.

Since then the area of woodland has increased, but forestry has continued to be the poor economic relation to agriculture and still tends to be concentrated on sites that are marginal for farming, for example on those that are steep or poorly drained and too difficult to farm, or those with poor soils and low fertility known as Less Favoured Areas (LFAs). However, whilst the income from timber may be limited in comparison to farming, woodlands can also offer many other benefits, such as providing important habitat for wildlife, adding to the amenity of the countryside, assisting with flood alleviation schemes and the improvement of water and air quality. Until recently these qualities, now often referred to as ecosystem services, have not provided most owners with any tangible financial returns, but this situation is already starting to change. There is now an increasing recognition of these different non-timber benefits that woodlands can provide to not only the owner but also the local community and wider society as a whole. This in turn is providing increasing opportunities for landowners to derive additional income streams from these benefits through both government grants and free-market mechanisms. It is already possible to sell the carbon predicted to be sequestered by a new woodland creation scheme, and there are opportunities to generate additional income from biodiversity offsetting, for example, to help facilitate the granting of planning consent for a new building development project. In the future, the opportunities to derive income from the range of public benefits that woodlands can provide, is expected to increase.

As a result, forestry is increasingly being seen as a financially advantageous land-use option, particularly for the areas of poorer-quality agricultural land. Thus, there is now an increased interest in planting land with trees. There is of course a range of important factors to take into account when considering such a land-use change. The starting point should be to consider what one hopes to gain from planting the land with trees. The motivations may be primarily financial, in which case one wants to assess the financial returns of continuing to farm as opposed to planting the land with trees, both now and what one might expect in the future. The impact on land values and taxation are also important considerations. There may also be other 'drivers' for wanting to plant land with trees, such as enhancing the landscape or habitat, or improving the sporting potential of the land. Finally it is important to remember that planting trees is a long-term venture.

Figure 1. A new woodland-creation scheme in an agricultural field adjoining existing woodland. All the new trees are protected in shelters or spiral guards, and there is a weed-free zone around each one to encourage successful establishment.

Under the current felling regulations it constitutes in effect an irreversible land use as permission is generally required to fell trees, and one of the usual conditions of consent being granted is the requirement to restock the area with more trees.

TYPES AND CATEGORIES OF WOODLANDS AND OWNERSHIP OBJECTIVES

There are many different types of woodland in terms of species composition, age, structure, condition and size. There is also a diverse range of both private and public-sector owners. The private sector includes not only farmers and landowners but also private investors, forest-management companies and 'lifestyle' buyers. Public owners currently include the Forestry Commission, Ministry of Defence, Local Government such as the county, unitary and local councils. Charities, such as the Woodland Trust, National Trust and the Wildlife Trusts, also own significant areas of woodland. Commensurate with this diversity of woodlands and ownership is the variety of owners' motives. Just like tree planting, reasons for managing existing woodlands can be financially orientated, such as seeking to derive the maximum financial return or to enhance the value of the investment, while for others, conservation may be a greater priority, or it might even involve a combination of motives. The owner's objectives will influence the forestry operations undertaken and in turn the composition of tree species, structure and condition of the woodland.

In addition to the type of woodland and the owners' intentions, the layout, distribution and relationship with other land uses, particularly within the same ownership, may also have an important effect. For example, woodlands may form an integral part of a farm or rural estate, or they may represent distinct entities in their own right. These factors too are likely to affect both the owner's objectives and will influence the forestry operations undertaken and the management strategy.

Where woodlands form an integral part of a farm or an estate, their main significance may lie in the benefit they offer to adjacent and nearby land and properties. For instance, a small block of woodland may add considerably to the amenity and sporting value of an estate, and thus the benefit it confers to the business overall is significantly greater than in its own right.

The majority of these woods on farms and estates tend to comprise smaller blocks of woodland and are in the form of spinneys and shelter-belts. Often they are also located on areas of land that are less suited for productive agricultural use, and may have been planted specifically to protect against wind and extreme weather, to provide shelter in order to prevent soil erosion or to improve amenity or sporting aspects, rather than for the production of timber.

Numerous studies and reports indicate that much of this resource is not actively managed and that the timber contained in it is often of low quality. The opportunities to derive any significant income from the production of timber will therefore be limited by the lack of good quality timber.

There are also woodlands that form distinct entities. The more commercially orientated woodlands tend to comprise large areas and are dominated by coniferous species, in particular spruce, pine, larch and fir, which grow considerably more quickly than most broadleaved species. Certain conifer species can better tolerate poorer sites too, such as infertile upland hillsides. In turn, these woodlands tend to be managed with the main objective of maximizing timber production. So the quantity and quality of the timber that these woods contain and the logistics and costs

Figure 2. Trees and woodland can shape a landscape, and thus care and thought is needed when planning and designing areas of woodland to be felled and restored. Felling can also enable previously hidden views to be exposed, as is the case here.

of harvesting the trees are likely to be the key considerations. By contrast the main impetus for the management of smaller blocks of broadleaved or mixed broadleaf and coniferous woodland has tended to be on biodiversity, amenity and field sports, and less about timber production.

THE FOREST RESOURCE

Physical site characteristics such as soil type, depth and drainage, as well as climate, elevation and the threat of wind damage affect the range of species that can be grown successfully. For example, most tree species can be grown successfully on a sheltered site with a brown-earth soil, whereas for an exposed upland site with a poorly drained soil, the choice of species will be more limited. The UK has a wide range of site conditions, but the region as a whole has the scope to support a very extensive range of species. Indeed, it has a rich heritage of trees that vary not only in growth and utility, but also beauty and interest. However, while there is an extensive diversity of tree species planted in gardens, parks and arboreta, the UK's woodlands are dominated by just a few key species. The conifer woodlands are planted mainly with four species: spruce, fir, pine and larch, as they have the capacity to grow quickly and produce good quality and marketable timber if located on the right sites with appropriate management. The broadleaved trees, too, are dominated by very few species: the main ones are oak, birch, ash, beech, and sycamore, which between them, account for over 80 per cent of broadleaves planted. While this strategy of focusing on a limited number of species has certain advantages, such as ease of management, economies of scale and concentration of efforts on what have been seen as the most successful varieties, there are also significant risks, in particular from pests and diseases.

Tree health has always been an important consideration, but over the last ten years it has increasingly become a more pertinent issue. Diseases such as 'sudden oak death' (*Phytophthora ramorum)* and 'ash dieback' (*Hymenoscyphus fraxineus;* also known as *Chalara fraxinea*) have highlighted the significant threats posed to particular types of tree, and thus underline the desirability of having a greater diversity of species.

The ash dieback disease, for example, has been particularly damaging to the main species of ash grown in the UK, spreading rapidly and killing infected trees. Few trees were affected up to 2012, but now the disease is found extensively throughout the UK.

Figure 3. Ash trees suffering from the chalara ash dieback disease. Much of the tree in the foreground, including the tips of the branches that have now died back, is typical of the more advanced stages of infection from the disease.

There are, of course, many other diseases and pests that can cause harm to groups and sometimes whole areas of trees, which can damage both the economic performance and the amenity appeal of the woodland. They are too numerous to detail here, but further information can be gained from sources such as the Forestry Commission and the Royal Forestry Society. While many of these different pests and diseases may not be as devastating as sudden oak death and ash dieback, they can still eventually kill a tree, and in the intervening period reduce its vigour and productivity. These effects make a strong argument for increasing the diversity of tree species grown commercially. Moreover, if a wood is made up of more than one species, the impact of diseases such as those mentioned are less devastating. Indeed, there might be enough trees still to have a viable wood, whereas if it comprised just one species, it could be necessary to fell the whole wood and start again. Furthermore, increasing the diversity of the tree species grown can also be beneficial for biodiversity and amenity, as well as helping to make a woodland more resilient to the impacts of climate change.

Many of the current models and thinking about climate change are predicting warmer, wetter winters and hotter, drier summers for the UK, as well as an increase in more extreme weather conditions such as gales, flooding and droughts. As a result, organizations such as the Forestry Commission are advocating that we should be growing trees that are better suited to these anticipated conditions. There are other people, of course, who question these predictions, with some suggesting that the climate in the UK may in fact become colder over the next 100 years. There is also some uncertainty as to how, and to what extent, trees will adapt to changing climatic conditions. Predicting something as complex as the weather over the lifespan of a tree, which may be over a period of 100 years or more, is undeniably difficult and fraught with uncertainty, and so is inevitably open to challenge. Yet, what is clear is the desirability of matching the right species to the site, diversifying the range of tree species grown and varying the age and structure of the woodland.

FORESTRY POLICY AND FISCAL INCENTIVES

Since the formation of the Forestry Commission in 1919, successive UK governments have sought to encourage tree planting and the appropriate management of woodlands. To help achieve these aims, fiscal incentives in the form of both grants and tax concessions have been offered to private owners.

These basic principles are still promoted today. Initially the concentration was on timber production, whereas now the focus is on multi-purpose forestry, including the provision of recreation and access, enhancing the landscape and protecting the environment. There is also recognition of the wider benefits that trees and woodlands can make towards local communities and society as a whole, for example in the form of improved water and air quality and increased public health and well-being.

Whilst all four countries of the UK have these same basic aims, following devolution England, Scotland, Wales and Northern Ireland have all developed their own policies and strategies for forestry. While there is still much in common between each country, differences in emphasis have been emerging as national administrations seek to reflect their own priorities. In turn, each country has its own schemes for grants. This section only gives an indication of the types of grant available in England, but further information can be gained from sources such as the Forestry Commission. At the time of writing, grants and other incentives are currently being assessed as part of a wider review of the UK's departure from the European Union.

Grants

Grant-aid has been an important strategic instrument for implementing forestry in private woodlands. A range of incentives has been available to encourage landowners to plant their land with trees, though more recently these have been targeted according to location and type of woodland, reflecting the priorities of the respective countries. Tree-planting schemes located close to large centres of population, with the provision of public access, as well as those designed to enhance the biodiversity of an area or for flood alleviation, have been the ones most likely to receive grants. These grants have included capital grants, designed to offset a significant proportion of the planting costs and for farmland. There also have been annual payments to help compensate for the loss of income from agricultural production. Despite these incentives, the relatively low returns from the production of timber, and the long-term nature of forestry compared with agriculture, has resulted in relatively small areas of land being converted to forestry over the last 20 years. Nevertheless, there is an expectation that there will be much more land planted with trees in the future. The UK Government has recently signalled its wish to see significantly more land planted with trees, setting a target of 30,000 hectares of new woodland creation each year until 2025. As a result, it is expected that incentives for new planting will not just continue, but also increase. The drive to encourage the 'right type' of tree in the 'right location' for the 'right reasons' is also expected to continue.

There is also a range of grants to help encourage sustainable management, and like those for new planting these are targeted. The grants for existing woodland are increasingly focused on semi-natural mature broadleaved woodlands, particularly ones deemed to be of high environmental value, and are used to encourage works designed to enhance the public benefits of woods, such as the improvement of environmental attributes or enhancing public access.

Grant-aid has helped fund the costs of several woodland managerial operations such as upgrading existing tracks or putting in new ones for access, fencing, deer management and clearance of non-native and invasive species such as rhododendron and Japanese knotweed. Further information on the grants available can be gained from sources such as the Forestry Commission website.

Tax

Woodland owners, historically and currently, have received benefit from several tax concessions: these include potential relief from the two capital taxes concerned with inheritance and capital gains. Relief from these two capital taxes will depend on whether the woodland is considered to be 'commercial' or may rank as part of the overall agricultural property.

If the woodlands are regarded as 'commercial', they are potentially relieved from inheritance tax, as they are currently able to qualify for 100 per cent Business Property Relief. They may also benefit from relief from Capital Gains Tax, where the value of the standing trees will be excluded from any gain. On the other hand, neither exemption would be available if the woodlands are not regarded as 'commercial'. However, the definition of 'commercial' within the legislation for both capital taxes is far from definitive. The basic premise is that the woodlands should be managed on a commercial basis with a view to the realization of profits. Thus, if the woodland has not been actively managed for a number of years, and little or no income has been derived, it may not be considered as being managed 'commercially' and so may not qualify for Business Property Relief.

The sale of timber is also potentially free from income taxes; as from April 1993, commercial woodlands were removed entirely from the scope of income and corporation tax. Despite this, such tax treatment is not always beneficial because expenditure cannot be allowed as a tax deduction against other income. So, for example, it is not possible to offset the often considerable expenditure on restocking and maintaining trees in their early years, thereby reducing the motivation to undertake these important works.

ASSESSMENT OF WOODLANDS AND THEIR MANAGEMENT

Woodland is a diverse and often complex resource and the ability to grow timber successfully, as well as provide a range of non-timber benefits, will depend on several factors. These elements include the physical characteristics of sites, including legal and regulatory issues such as rights of access, occupation, environmental designations and constraints, as well as the tree crop itself. The management strategy and subsequent operations undertaken are also important in determining the economic, environmental and social benefits that a wood can provide.

There are a number of different management operations undertaken by woodland owners and their managers. These include restocking of woodland by both planting and natural means, the weeding and maintenance of these newly established trees, thinning and felling, and pest control. The choice, frequency and extent to which these different operations

Figure 4. Mixed broadleaved woodland. Small coupe-felling has recently taken place to diversify the age structure of the woodland and been restocked with more broadleaved trees, planted in plastic tree shelters to protect them from rabbits and deer.

are undertaken will depend on the age and condition of the woodland, the physical characteristics of the site, forestry policy and economics and the management objectives.

Restocking of woodland

The restocking of a woodland area is usually undertaken following the felling of some or all of the trees. It may also be done following natural events such as storm damage or loss of trees through pests and disease, or to add diversity to the woodland. Restocking can be achieved both by the planting of young trees and by natural regeneration from the seed of existing trees. Sometimes both approaches are used. Whichever restocking method is followed will depend on several factors including the tree species present on the site and those desired in the future, in line with the manager's objectives. Natural regeneration is less predictable in terms of future species composition, stocking density and timing, but can create a more natural-looking wood at a lower cost than by planting. Newly established trees will be at risk from the ravages of rabbits, hares and deer. Thus it is important to consider whether such a threat is present and, if so, to protect the trees. This can be done by fencing the area or by individual tree guards. Alternatively it may be possible to control the population of these animal pests. Newly established trees require regular inspection and maintenance, including 'beating up' and weed control. Beating up is the procedure carried out to replace any losses of newly planted trees that might occur during the first couple of years.

Weed control is usually carried out for the first three to four years and can be controlled by herbicides, hand and mechanized weeding and mulches. Weeds compete for light and nutrients, and so have a significant impact on the height, growth, general vigour and survival of the newly established trees. However, weed control is costly and may involve herbicides that have some environmental drawbacks or are restricted in their application. So for these and other reasons, it is suggested that weeding should only be done when considered necessary and beneficial.

It may also be necessary to clear unwanted trees or shrubby growth, such as rhododendron and elder, that is competing with the newly established trees. It may also be desirable to reduce the stocking density of the newly established trees. This is particularly the case when restocking by natural regeneration, which can lead to high plant density. In such cases clearing or 're-spacing', to which it is sometimes referred, is undertaken to remove unwanted species and crowded trees to achieve the desired spacing.

Thinning

The key silvicultural activity once establishment is successful is a procedure known as 'thinning'. A first thinning of the trees may start as early as 12 to 15 years old after planting, particularly on better sites with faster-growing species. At the other extreme, slower-growing trees on poorer land may not have a first thinning done until 30 years old. Thinning involves the removal of a proportion of the trees in a wood to satisfy both economic and environmental objectives. Trees that are allowed to grow close together for too long produce thin, weak stems with narrow crowns and restricted root-systems.

If thinning is delayed too long, the trees become suppressed and can die resulting in economic loss, and those which do

Figure 5. This conifer woodland has been well thinned over the years to allow sufficient light to reach the woodland floor and promote natural regeneration and other ground vegetation.

survive are likely to be less valuable, particularly as their girths tend to be restricted. So where the production of top-class timber is considered to be an important objective, the poorer-quality trees are likely to be removed through a number of thinning operations. This is commonly known as 'selective thinning', which aims to concentrate the growth on those trees most likely to produce high-quality timber by giving them more space and reducing the competition for light and nutrients. Other benefits include financial return, assuming that the timber price exceeds the harvesting costs. It can also enhance the habitat and amenity values of the woodland, as a result of the increased amounts of light reaching the woodland floor, which generally encourages a greater variety and amount of ground and shrub vegetation. Thinning can also help improve the health and vigour of the retained trees.

Figure 6. A forwarder loading felled timber from a thinning operation.

The amount of timber removed at each thinning depends on several factors, including the species, site fertility and in turn the growth-rate and age of the crop, the state of the timber markets, the economics and the management objectives. As a very rough guide, about 15 to 30 per cent of the trees may be removed at a thinning. In terms of frequency, younger woodlands need thinning more often than older ones, and the faster-growing conifers more often than the slower-growing broadleaves. For conifers, thinning might take place every four to five years, extending to up to every ten years as the trees get closer to final felling, whereas broadleaves tend to have longer thinning cycles of every seven to ten years to start with, extending up to every 15 years as the trees mature.

Figure 7. A timber harvester working in a mature conifer plantation. The harvester not only fells the trees, but also removes their branches and tops. The harvester's operator then cuts the timber in predetermined lengths, seen in the foreground, to be collected by a forwarder and taken to a stacking area for onward transport to end-users.

Felling and the optimum rotation length

The time between planting and final felling is called 'rotation', and this may vary from as little as 20 years for fast-growing poplars to more than 200 years for oaks. But what constitutes the optimum length of rotation? There are various factors that can influence this cycle. These include the tree species and their growth-rate and expected lifespan, the market conditions and the owner's objectives. Different trees mature at different ages. The rate of growth depends not only on species and site factors but also on the age of the trees. Once the trees reach maturity, their growth-rate starts to slow significantly. Thus, if timber production is an important consideration, there are likely to be pressures to fell the crop at or before this point. Likewise, the state of the market and whether the timber has reached its optimum marketable size are also important reasons for doing so. By contrast, where habitat and landscape quality are the primary objectives, there may be strong reasons for retaining the trees beyond their economic rotation.

As well as considering the length of time for rotation, it is also important to consider the scale and type of felling. Even-aged woodlands are often felled in their entirety ('clear-felled') or in a larger section at the end of their rotation (see Figure 7). However, clear-felling creates a sudden and dramatic change in the landscape, and breaks the continuity of woodland conditions. Thus such an approach is likely to be regarded as inappropriate for woodlands having a high value for habitat, landscape or even sport or recreation. In these cases other approaches should be considered by using what is often referred to as a 'continuous-cover' system. This involves small-scale felling of 'coupes' down to removing individual trees or small groups of trees (see Figure 4). By adopting such an approach, it is possible to maintain a continuously wooded area. These systems are more complex

Figure 8. Felled timber from a thinning operation, stacked at the edge of a forest track and ready to be transported to the sawmill. The stacked timber here is of Western red cedar and has been cut into lengths specified by the end-user. Western red cedar has a low density and low strength, but has good resistance to fungi and so tends to be used for external weather boarding and fencing.

and time-consuming and to date are not widely adopted in the UK. However, they are attracting increasing interest, particularly as the concern about the environment and the potential impact of climate-change increases. A detailed description of these silvicultural systems is outside the scope of this book, but further information can be gained from various sources, including the Forestry Commission and the Continuous Cover Forestry Group.

It is also important to be aware that the felling of trees, both from thinning and clear-felling operations and the subsequent extraction of that timber can cause damage to the soil and ground vegetation. So the timing of such operations, as well as the choice of the harvesting system and contractors used are also important considerations, to help minimise any damage to the site – this is particularly important on sensitive sites with soils that are prone to compaction. Furthermore it is also important to try to minimise disruption to other activities being undertaken in the wood, such as a pheasant shoot.

Other management operations

Other important management operations include the control of animal pests, in particular deer and squirrels. For example, squirrels can strip the bark from trees; broadleaved species such as beech, sycamore and oak that are aged between 10 to 40 years are particularly vulnerable. Deer and rabbits pose a significant threat to newly established trees and may also cause damage to the ground flora. Managing their populations at sustainable levels therefore may be desirable. In cases where habitat and sporting objectives are important, the creation of rides and glades, the retention of dead wood and over-mature trees for the benefit of insects, species of birds and other wildlife may also be desirable. Finally general maintenance is usually advisable, which would include the upkeep of fences, watercourses, drains, access routes, rides and boundaries.

TIMBER MARKETS AND THE ECONOMICS OF WOODLAND MANAGEMENT

As mentioned above, woodlands are managed to produce both timber and non-timber products. There is also a range of factors that will influence the economics of producing these different benefits. Despite the increased opportunities to derive income from the non-timber benefits, the production and sale of timber is still a key reason for owning and managing many woodlands. Therefore the section below is focused on the sale and marketing of timber and the factors affecting timber prices and returns.

Timber markets and prices

The main markets for timber have traditionally been, and still are, the wood-processing industries which are based on four main forest-product sections: sawmilling, paper and board, wood-based panels and fencing which includes other wood products such as garden furniture. The prices paid by these timber-processing industries will depend on the diameter, quality and species of trees that in turn will influence the products into which they can be converted. There are also essential differences between the markets for coniferous and broadleaved trees.

For conifers, the thinner timber of 7 to 14cm, known in the trade as 'small diameter timber', is generally sold to the wood-based panel industries, in particular for chipboard. Spruce, in particular, as a 'white wood' is also sold to the pulp markets. The larger, round wood, with a diameter of over 14cm, may be sold to the fencing plants and sawmills. Notwithstanding the above, the market for wood fuel is now an increasingly important outlet as a result of the increased adoption of biomass heating systems, and the increased use of biomass by the UK's power stations, to generate their electricity.

The main markets for broadleaved timber are found in sawmills, so long as the timber is straight and free of defects and with sufficient diameters – generally over 40cm. Otherwise, the main market currently is for firewood. As with biomass, the demand for broadleaved-timber firewood has increased significantly over the last 20 years. Timber from species such as oak and sweet chestnut is also used as fencing material.

Figure 9. Timber stacked in the yard of an estate sawmill. This timber 'in the round' will be subsequently sawn and processed into a range of products including for use in the building construction sector

There are also several 'niche' markets for both broadleaved and coniferous timber, including transmission poles, turnery, mulch and animal bedding.

The state of the timber market will also affect the prices offered. Timber and wood products are subject to the global free market, and so British-grown produce must compete in terms of price, quality and service with imported timber. The most important factors that affect world prices are supply and demand, as well as the costs of production and prices of substitute markets for wood. The availability of supplies of timber from countries such as Sweden, Finland and the Baltic states, together with the relative strength of sterling against the currencies of those countries, also influence timber prices.

Thinning and felling are the main forestry operations that are most likely to result in the production of saleable wood. This timber can either be sold as standing trees, where the buyer is responsible for felling and extracting them, or it can be felled and left at stump, extracted to the roadside or delivered to the final user, which might be the sawmills and the wood-using industries. The main factors that affect the returns will be the value of the timber and the cost of harvesting and delivery to the end-markets.

ISSUES FOR LAND MANAGERS — A SUMMARY

As can be seen in the text above, there are many important issues that land managers need to consider when managing woodlands. These include the more dramatic decisions such as when and how much of the woodland to fell and the method of restocking, be it relying on natural regeneration or planting with new trees. In turn, the choice of tree species is becoming an increasingly important consideration. While the threats posed by pests and diseases may be the most compelling reason to further diversify the species grown, there are other reasons which include biodiversity, amenity and climate change.

Appropriate management and maintenance of a woodland is also important: the regularity and intensity of the thinning operations will influence the quality and quantity of the timber grown, the health and vigour of the crop, as well as the type and extent of ground vegetation within the woodland, which will have an impact on its habitat and sporting value. The timing of thinning and felling operations, as well as the choice of the harvesting system and contractors used, are also important considerations, to help minimise any damage to the site or disruption to other activities that might be taking place within the wood. Other operations, such as squirrel and deer control, may also influence the quality of both the timber crop and the habitat value, and so also need to be considered. Not to be forgotten are policy and fiscal considerations, such as the availability of grants and tax incentives that will also influence the type and extent of the management undertaken.

The starting point for any land or woodland manager is to get to know the woodland and to really understand what the owner wants from his or her woods. Some will have financially orientated motives including enhancing the value of their investment, while for others it may be conservation or amenity, or a combination of motives. It is only when you understand these points that you can really determine how best to manage a woodland.

CONCLUSION

Forestry as an alternative land-use option, in particular whether or not to plant areas of poorer-quality land with trees, is likely to be an increasingly important consideration for rural land managers in the future. There is a wide range of factors to consider when contemplating such a land-use change. These include motivations for wanting to do so. They may be primarily financial, in which case one wants to assess the financial returns of continuing to farm as opposed to planting the land with trees. The impact on land values and taxation are also important to understand. There may of course be other 'drivers' for wanting to plant land with trees, such as enhancing the landscape or habitat, or improving the sporting potential of the land. Finally, it is important to remember that planting trees is a long-term venture and under the current felling regulations it can in effect amount to an irreversible land use. Whatever the decision, the old mantra of the 'right tree for the right place and for the right reason' is likely to be as important as ever.

FURTHER READING

Starr, C. (2013). *Woodland Management – a Practical Guide.* Crowood, Ramsbury.

Forestry Commission publications, Edinburgh:
(2017). *UK Forestry Standards.*
(2015). *Thinning Control - Field Guide.*

Index

About the Author

David Lewis
Dip Est Man, MPhil, MRICS, FAAV, MICFor, FHEA

David has worked in education and land and woodland management in the United Kingdom for over 25 years. At the start of his career, he worked for the Prudential Insurance Group, assisting with the management of their rural-investment property portfolio. He then went into education. He is currently a senior lecturer at the Royal Agricultural University (RAU), Cirencester, and is the course manager of the University's MSc in Rural Estate Management. He also lectures on Woodland Management and Rural Asset Management, and is a qualified Chartered Surveyor, Agricultural Valuer and Chartered Forester. His MPhil focused on assessing the feasibility of using wood, including energy crops, for the generation of energy.

David's research and consultancy mainly focuses on the management of farm and rural-estate woodlands, forestry policy, economics and valuation. He has accrued over 25 years of experience in forest management, and is currently involved in the maintenance of approximately 1,000 acres of lowland woodland, including significant areas of SSSI and ancient woodlands. He has recently been asked by the Royal Institution of Chartered Surveyors to represent their members on the Forestry Commission's Applicant Focus Group to provide industry feedback, prior to the launch of new policies and initiatives, including new schemes for grants and measures of support.

The manufacturer's authorised representative in the EU for product safety is: Easy Access System Europe, Mustamäe tee 50, 10621 Tallinn, Estonia, https://easproject.com gpsr.request@easproject.com

Printed and bound by CPI Group (UK) Ltd, Croydon, CR0 4YY

27/02/2026

02062027-0001